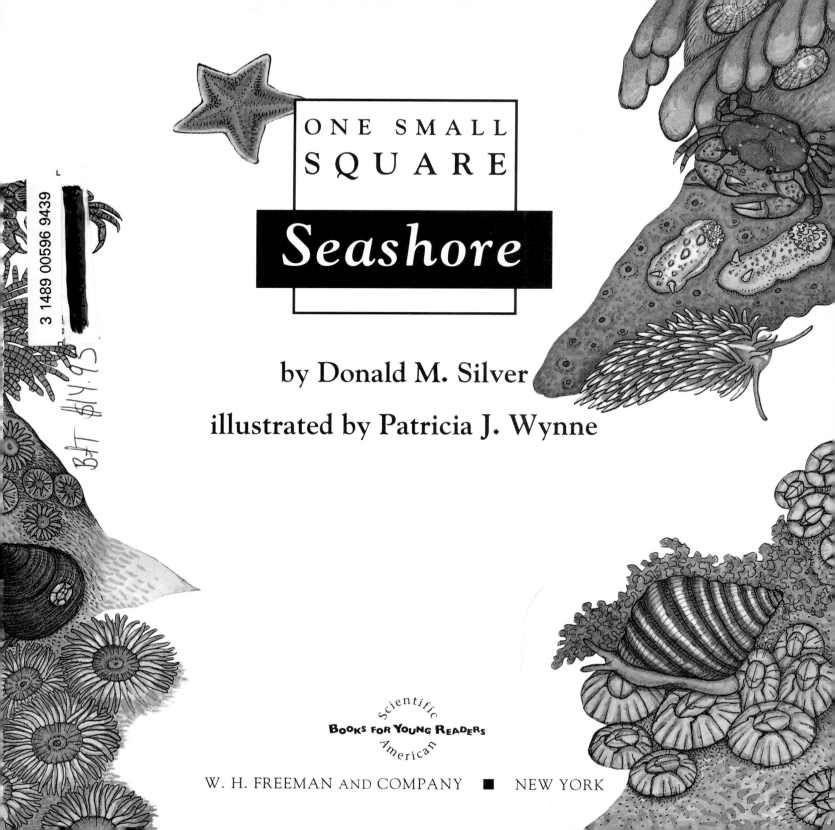

ONE SMALL SQUARE

Seashore

by Donald M. Silver

illustrated by Patricia J. Wynne

Scientific
BOOKS FOR YOUNG READERS
American

W. H. FREEMAN AND COMPANY ■ NEW YORK

Every plant and animal pictured in this book can be found with its name on pages 36-43.
If you come to a word you don't know or can't pronounce, look for it on pages 44-47.
The small diagram of a square on some pages shows the distance above or below the
ground or the water for that section of the book.

Dedication

For my parents
—who allowed me to play in the mud.
Patricia Jewell Wynne

Thanks to

Bonnie Cathey, Wini Manoian,
Tillie Beavers, and Thomas Cathey
for the many hours spent in search
of shore life and shells along the
sandy beaches of Sanibel and
Captiva islands, Florida. My grateful
appreciation to Ivy Sky Rutzky for
her assistance and advice.

Text copyright © 1993 Donald M. Silver.
Illustrations copyright © 1993 Patricia J. Wynne.
All rights reserved.

Printed in the United States of America

Scientific American Books for Young Readers is an imprint of W. H. Freeman and
Company, 41 Madison Avenue, New York, New York 10010.

Library of Congress Cataloging Number 93-18354
ISBN 0-7167-6511-X
10 9 8 7 6 5 4 3 2 1

Introduction

Something is always moving at the seashore. The tide comes in and goes out. Waves break; water sprays; sand shifts; seaweeds sway; breezes blow; birds scurry; crabs crawl; clams dig; snails slide; and worms burrow.

To people swimming, surfing, jogging, and collecting shells, no place is more fun and relaxing. But to the animals and plants that live there, the shore is not a vacation spot. It is home at the edge of the sea, one of the most difficult places on Earth to "make a living."

The sea covers almost three quarters of Earth's surface, surrounding all continents and ocean islands. Everywhere land and seawater meet, there is a seashore. Some shores are sandy beaches that slope gently into the sea. Others are steep, rocky cliffs pounded by the full force of crashing waves. This book explores both.

3

When you visit a seashore, look around for signs of life. Most people see only water, seaweeds, shells, and birds—and think that is all there is. But there may be hundreds, perhaps thousands, of kinds of living things just waiting to be discovered.

Tools such as the ones on this page can help you dig, sift, and collect the things you find. Ask an adult if you may dig. Take along a notebook and a pen or pencil. Write down clues and draw shore creatures as they work to stay alive. Your notebook will help you remember what you saw, and you can add to it when you visit other places.

Watch the water at different times of the day. It seems as if the water in the ocean is moving wave by wave to shore. Waves are really not moving water but moving energy. (See illustrations on pages 6 and 7.) Waves form when winds blow across the surface of the ocean and give some of their energy to the water. The water moves up and down, but only the energy moves forward. When a wave reaches shore, the water moving up and down scrapes the shallow bottom, curls, and rolls over. It breaks into foamy surf that rushes ashore.

Use an inexpensive magnifying glass to discover tiny shore creatures, clues to where they live, and dazzling patterns on the smallest shells washed up on the sand.

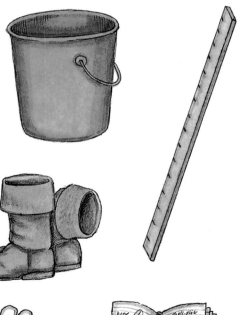

A small shovel or gloves for digging, a piece of screen and a pan for sifting, and a pail for collecting are among the best tools for the seashore explorer.

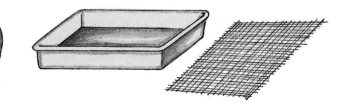

5

The moon circles the Earth about once a month. Each day as Earth turns, the moon's gravity pulls on the ocean and the land. When the ocean faces the moon, the water bulges out away from the land. It is low tide on shore.

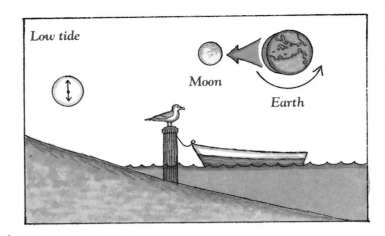

Low tide

Moon

Earth

When the land faces the moon, seawater is pulled toward shore. It is high tide. Every day along most ocean shores, water rises for about six hours, falls for about six hours, then rises again in a never-ending cycle. The part of the shore between high tide and low tide is called the intertidal zone.

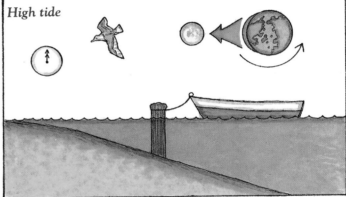

High tide

Watch a gull bobbing up and down on the water. As each wave moves forward, the gull stays in the same spot, because the water isn't moving. Only the energy in the wave moves. Try making waves in a bathtub and watching a floating object move up and down.

As a wave nears shore, it scrapes bottom, slows down, curls, and breaks into the foamy sheet of surf that washes up on land. Waves that bring sand to shore build beaches. Waves that pull sand away erode—wash away—beaches. Storm waves can destroy a beach in hours.

Although waves don't change the overall water level, they do shape and reshape shorelines by carrying sand to and from beaches and pounding away at rocky cliffs.

There is something that does make the actual water level on shore rise and fall. This rise and fall is called the tide. The tide is caused by the pull of the moon's gravity on Earth as Earth spins around completely on its axis once every 24 hours. When the moon's gravity pulls seawater toward land, the water level rises—until at high tide much of the shore is covered by saltwater. When the moon pulls water away from land, the level falls—until at low tide most of the shore is exposed to the air. Any plant or animal that lives on the part of the shore covered by high tide and exposed by low tide must be able to survive both in and out of seawater.

Look at a handful of sand under your magnifying glass. If the grains are round and polished, they are most likely the mineral quartz that came from rocks broken down thousands of miles away, picked up by rivers, and carried to the sea.

Pebbles

Igneous rock

Metamorphic rock

Sedimentary rock

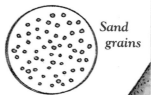
Granules

Sand grains

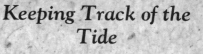

Keeping Track of the Tide

On most seashores there are two high tides and two low tides each day. But some shores have just one high and one low tide. *The best time to explore the shore is when the tide is going out.* Local newspapers, weather channels, and libraries have daily tide tables that list the times of high and low tide in the area of the shore you are visiting. Check to see if the local table is correct for your shore by keeping your own tide table.

where	day	high	low
Battery	Feb 24	10:22 am	4:30 pm

The shape of the land and the depth of the water can affect the times of high and low tide. Even so, you will soon notice that each day the tides occur about 50 minutes later than on the previous day. Each day, the moon moves a little in its orbit. It takes Earth about 50 minutes to catch up.

7

One Small Square of Sandy Shore

The swimming ostracod is the size of a pinhead.

With speed and skill, the jackknife clam uses its strong muscular foot to dig down into the sand.

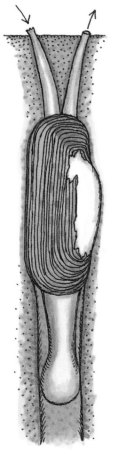

When hunger strikes, the rove beetle takes off after sand hoppers on the upper beach.

Look along the beach for a wavy line of shells, seaweeds, and driftwood. This is the strandline, the farthest water reaches at high tide. Take a yardstick or a meterstick (about three inches longer) and with your finger draw a square in the sand the length of the stick on each side. Be sure the strandline is inside your square as it is in the illustration. The low tide waterline is in the square in this book. Only if the beach is narrow will it be inside your square too.

If you fully explore your square, you will discover what makes the seashore so different from any other part of nature. For instance, just about anywhere else you could draw your square and return day after day to find your lines still there. But because most of your square will be underwater at high tide, you will have to keep redrawing the part of your square that lies below the strandline and is washed away.

Follow along as a square of sandy beach and then a square of rocky shore are explored. There will be activities that you can do in your square. While you may not find all of the living things in this book on the seashore you visit, you will find similar creatures wherever the land meets the sea.

The sand hopper hunts for tiny animals between grains of sand.

Not all sandy beaches are alike. One small square on a narrow beach may reach from the shallow water to above the strandline, where the sea doesn't go.

You would need a microscope to see the tiny plants (left) and animals (right) that make up the plankton that floats in seawater.

Whether swimming, digging, or crawling, the inch-long mole crab always moves backward.

The red-winged blackbird warning other birds to keep their distance doesn't bother the piping plover protecting her young.

American beach grass

Red-winged blackbird

Ghost crab

Raccoon

Piping plover

Plovers run up and down a sandy beach searching for food. Above the strandline, they hunt insects, sand hoppers, and pill bugs crawling in grains of sand. Plovers nest on the beach, so watch where you step.

Skunk

Mole cricket

Fox

Blown around by a breeze, a broken blade of American beach grass draws a sand circle. Never pick shore plants. Their roots hold sand in place, helping to stop erosion.

Check for tracks left by animals that visited your square at night. Look for scratches of ghost crabs; dents of sand hoppers; and footprints of raccoons, skunks, foxes, and birds.

10

The Beach Ghost

If you visit your small square early in the morning, you might see a ghost—a ghost crab, that is. Keep your eyes on it. Its color so closely matches the sand that the crab seems to appear out of nowhere and disappear again.

Along with pill bugs, sand hoppers, and a few kinds of plants, ghost crabs live above the strandline where seawater doesn't reach. Follow a ghost crab, and you may see it crawling back into its deep burrow after a night of searching for food. Watch how quickly the crab seals the burrow opening with sand to keep out wind, sun, and predators—animals that might eat it.

What's living on shore?

Plants

Animals

Funguses

Protists

Monera

Write down in your notebook anything and everything you discover at the shore. Record the date, time, and weather. Draw pictures of animals, plants, and shells you want to learn more about later.

To catch animals living in the upper beach sand, cover a pan with a piece of screen and slowly pour a shovelful of sand through it, shaking gently. With your magnifying glass, keep checking the pan for animals. Sketch each one in your notebook. Return the sand and animals to where you found them. Each kind of beach animal lives best where it is adapted—fitted—to survive.

The ruddy turnstone flicks over every stone and piece of seaweed it finds along the strandline. It captures its prey in its short beak.

The sand hopper burrows by day and feeds at night.

Stranded when the tide goes out, live clams, crabs, barnacles, and snails are attacked by shorebirds and other predators. Few will survive until the water returns.

Pill bug

Rolled up

Porcelain crab

Hermit crab

Sea anemone

Zebra flatworm

Marooned on the strandline of your small square may be a hermit crab that moved inside an empty snail shell. A porcelain crab and a zebra flatworm may share the same shell, feeding on food bits the hermit drops. In the water they would all be protected from predators by the stinging cells of the sea anemone that the hermit attached to the top of the shells. But on shore, the sting-ing cells are of little use against sun, wind, and predatory birds.

Semipalmated plover

Tiger beetle

Rove beetle

Gribble

Now that the tide is going out, the plover, tiger beetles, rove beetles, and earwigs can't wait to eat.

Earwig

Look carefully with your magnify-ing glass at shells to see if worms, snails, or other animals have tried to drill a hole in them.

Stranded

The tide is in. Most of your small square belongs to the ocean. All you can do is wait until the water pulls back. But the wait is always worth it, for you never can tell what the sea will leave along the strandline for you to find. And you are not alone. Also waiting for the latest delivery of gifts from the sea are hungry gulls, plovers, tiger beetles, and mole crickets. They are ready to search for tasty bits of plants and animals. At night, skunks and raccoons may come by to have a look at what has been cast up on shore.

One day it might be tangled seaweeds, broken shells, a piece of coral, or a dead fish. Next day it might be driftwood, sea urchin spines, and a black, leathery mermaid's purse—the egg case of a shark or a skate. Look for round holes and tunnels drilled into driftwood by shelled animals called shipworms (which are related to clams, not worms) or by gribbles, which resemble woodlice.

Lift seaweeds, wood, and shells, and with your magnifying glass try to catch a glimpse of tiny animals scurrying to safety. They must avoid being eaten by another shore creature that feeds along the strandline.

Portuguese man-of-war

Stinging Animals: Beware!
Anytime you find a jellyfish or a Portuguese man-of-war stranded on shore, leave it alone. These animals have stinging cells that can harm you even out of the water.

Moon jellyfish

Look What the Tide Brought In

Whenever you work in your small square, first check the strandline to see what the tide brought in and which shore animals are picking through the line when you are. Close your eyes and listen to the sounds that feeding birds make.

fox.

crab

raccoon

skunk

Keep It Clean

Not everything you find along the strandline should be there. Garbage, clumps of oil, tar balls, and broken glass are clues to the danger shores face because of pollution. Showing signs of pollution to your family and friends will help everyone remember not to litter or leave food behind. Food may attract predators that eat baby birds.

If you see needles or hospital wastes, don't touch them. Call an adult to report the hazard and ask to have it removed at once.

By hiding under a rock, the oval flatworm escapes the drying heat of the sun.

The Land Between the Tides

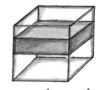

Sit above the strandline and watch your square as the tide goes out. Look for animals and plants that can live on the sand when it is exposed to the air. Many animals and plants tossed ashore by waves can't. Starfishes and sand dollars and snails still inside their colorful shells can breathe oxygen only if it is dissolved in seawater. The horseshoe crab that crawled into your square to lay her eggs in the sand must return to the sea or die. Seaweeds and other plants are stranded when the tide goes out.

Other animals don't live on the land between the tides for another reason. Neither the sanderlings darting in and out after each wave to fill their short bills

If you see slipper shells stacked on top of each other, it is likely that they contain live animals.

The Limulus leech lays eggs on the gills of the horseshoe crab.

The first horseshoe crabs lived on Earth millions of years ago. These "living fossils" are still laying eggs on sandy shores.

Dents in the sand can mean sand hoppers looking for food.

with food nor the gulls screaming and fighting over a tasty morsel live their lives there. Any eggs they laid would soon be washed away. The pill bugs, beetles, and other insects that cross the strandline in search of a meal are not safe here either. If they are pulled into the ocean, they will drown.

When your square is out of the water, use your magnifying glass to take a closer look at the surface of the sand. You may discover a sunken hole, a hole surrounded by a small pile of sand, a tube sticking out of the ground, or a tiny jet of water squirting out. These are all clues that the place to live between high and low tide is not on top of the sand but in it.

If you find a collar of sand on shore, it contains eggs laid by the moon snail. Eggs and sand are held together by hardened snail mucus.

The five-pointed pattern on the sand dollar is a clue that it is related to the starfish.

Collecting Shells

If you go to the shore, collect as many empty shells as you want. Note when and where you found your shells. Make shell tracings of those you can't take home. Within each tracing, mark down features such as patterns and colors. Later, use a field guide from home or from the library to identify your shells.

A field guide is a book with the names and pictures of animals or plants that live in different places. There are guides for birds, insects, flowers, and seashells. With so many kinds of shells, you may not be able to identify every one you collect or draw. However, the more you know about shells, the more likely you will be able to tell which ones are related. If you find a shell with a live animal inside, take it to the water's edge so it can be pulled back into the sea.

Like its relative the earthworm, a clam worm (below) has bristles sticking out of each section of its body that grip the sand as it tunnels through.

At high tide, some burrowing clams dig out to feed. Others poke tubes above the sand surface to pull in water, which contains their food.

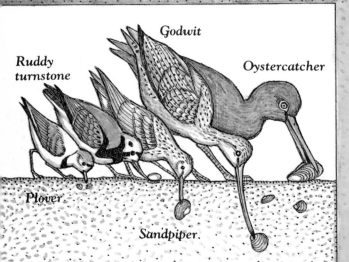

Godwit

Ruddy turnstone

Oystercatcher

Plover

Sandpiper

Each kind of bird has a beak of the right shape and size for reaching the kinds of shore animals it needs to eat.

In the midday sun the sand is hot and dry. But just a few inches below the surface, it is cool and wet. The film of water trapped between sand grains keeps the temperature about the same on the hottest and coldest of days. It prevents animals living there from drying up.

Within the sand there are diggers, tunnelers, and tube builders. Try finding them. As soon as they sense you digging, most of them dig even deeper.

Like customers in a restaurant, nearly all of these hidden creatures are just waiting to be served. When the tide comes in, so does dinner. The menu includes one-celled plants and animals, tiny copepods, and shrimplike krill. Clams open their shells and stick out tubes that pull water into their bodies. After they filter

16

In soft sand the pen shell attaches itself to a buried stone by strong threads called byssus.

When the milky ribbon worm (below) shoots its mouth tube at its prey, the barb at the end goes into the victim, which the worm swallows whole.

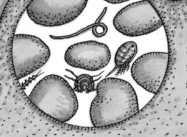

Swimming and crawling in the film of water trapped between sand grains are thousands of tiny animals.

the food from the water, they pump the water out. Snails and worms creep about searching for food bits sinking onto the sand.

High tide also brings fresh oxygen dissolved in seawater. As the tide pulls back, shelled animals trap water inside themselves so they can breathe until they are again underwater. Creatures that have come out of hiding to enjoy the feast quickly disappear from view.

Living buried in sand is no guarantee of safety. From above, there is the ever-present danger of a hungry bird. In the sand itself there are predators such as the moon snail. It can bore a hole in a clam shell, insert its mouth tube, and suck up the soft body of the helpless clam.

Digging In

At low tide, try to outsmart burrowers by digging beneath them and sifting the scooped-out sand on your screen. When you have sifted all the sand, check the screen for animals. Gently sift any animals you catch in your pan to look at with your glass and draw. Don't keep animals out of the sand for too long, or they will dry out and die. But you can replace animals on the surface and watch them dig to safety.

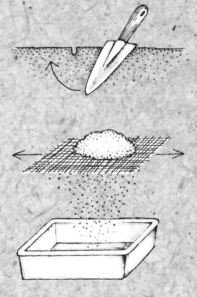

Sharing the Beach

Unless the beach is in your backyard, you will have to share it with many other people. Refill any hole you dig, so no one trips. At times, you may find someone else using the spot you chose. Then you will have to make another square.

Down by the Sea

Where the land ends, the sea begins. In the sea there is a vast world of water life spanning the globe. It reaches from shallow, sunlit waters at the edge of the shore to the frigid darkness of the midocean deep.

Countless animals live in the sea. Crabs, shrimp, sand dollars, sea biscuits, and starfishes live not far from your square. Sharks and other fishes wander in and out of coastal waters. Some creatures large and small never come close to land.

If the sea at low tide falls inside your square, take along an adult if you want to explore near the water. It

Colored black above and white below, a black skimmer is hard to spot for birds looking down or for fishes looking up.

Jellyfish tentacles are armed with stinging cells not only for capturing prey but also for keeping predators away.

The brown sea biscuit camouflages itself with sand to blend into its surroundings. The winter flounder can actually change color to hide itself.

Look for the gas-filled float of a by-the-wind sailor.

is impossible to observe sea life and also keep a careful eye on the sea itself. The adult can be on the lookout for waves that are strong enough to pull you in. Then you can watch a pelican dive-bombing after a fish, a skimmer flying inches above the surface and slicing the water with its open beak, or the pattern on a calico crab washed up at your feet.

One way or another, all the animals in your square of sandy beach depend on the sea to make their livings. The same is also true for the very different kinds of creatures struggling to survive on a rocky shore.

After blowing a raft of bubbles that the wind and waves will move, the purple snail clings upside down for the ride.

To escape a starfish, a scallop claps its shells together to squirt out a jet of water, and then it zigzags away.

Most of the time the angel shark lies almost buried in the sand, feeding on fishes and crabs. If it senses danger, it will swim to safety.

Wood, shells, worms, and other animals add to the decorator crab's disguise.

19

A Small Square of Rocky Shore

If you go from a sandy beach to a rocky seashore, just about everything changes. Of course, a rocky shore has waves, tides, drying winds, and scorching sun just like a sandy beach. But instead of a carpet of tiny grains of sand, there are mostly rocks—some boulders, others pebbles or even smaller stones. Instead of animals hidden from view, there are many creatures right out in the open in the middle of the day, even at low tide. There are plants growing on rocks that are covered by seawater when the tide is in.

Bands of color, sounds of smashing waves, the spray of water, large puddles, and slippery rocks are all signs that a rocky shore is very different from a sandy beach.

You have to be extra careful exploring a rocky shore. Look for a place where the waves are small and there aren't steep cliffs to climb. Ask an adult if the place you chose looks safe enough to be your square.

21

In addition to a notebook and pen or pencil, you'll need a piece of chalk and a yardstick, meterstick, or cloth tape measure to mark off your small square. Wear boots or sneakers for walking on slippery rocks. Take a net and a stick for lifting plants or poking around in tide pools.

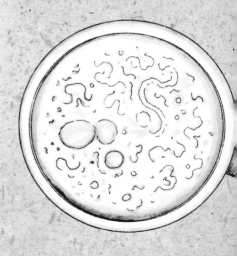

Have your glass ready to magnify slimy plants, tiny animals, hungry snails, and cone-shaped barnacles.

Don't mistake a sea anemone for a colorful flower. It is an animal armed with stinging tentacles.

A mussel's hard shell is a clue that many rocky shore animals survive sun, wind, waves, and predators.

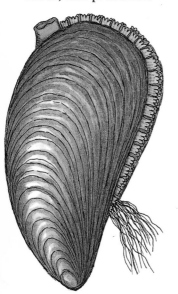

After a night of feeding on tiny plants, a limpet glides back to the same spot where it rests when the tide is out.

Rest spot

Always wear boots, sneakers, or rubber-soled shoes that can grip rocks made slippery by wet plants. Explore *only* when the tide is going out. You don't want to be caught on a rock when the tide comes in.

There is no strandline on a rocky shore. Instead there usually is a black band running along the rocks like the one shown in the picture. If there is no black band on a rocky shore you visit, look for a crust of plants growing on rocks back from the water. Include the black band or crusty rocks inside your square. They mark where living things are safe from high tide.

Because most of your square is underwater during high tide every day, mark off the uppermost part of your square by drawing a chalk line on the rocks. Bring your measuring stick or tape with you whenever you explore, to redraw the part of the square that is covered by the sea at high tide. If there is no tide pool in your square, explore one elsewhere on the shore, out of reach of pounding waves. (See page 28.)

The square of rocky shore in this book was chosen to be away from the water even at low tide. This is to remind you to keep an eye on the tide so that you will always be safe.

Sea slug

Boring sponge

Look carefully into a tide pool, and you may see a sea slug feeding on a boring sponge.

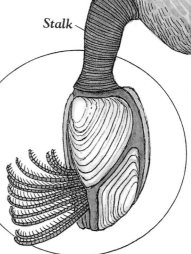

Stalk

Cemented to an overhanging rock by its long stalk, a goose barnacle extends its feathery feet to catch food when the tide covers it.

The tidepool sculpin deserves its name, because this fish is found in nearly every tide pool.

You may find a flowering plant such as thrift growing from a patch of soil in a rock crevice. It is one of the few kinds of plants that can live above the reach of high tide, in the salty sea spray of the rocky shore. Draw it in your notebook, along with any animals crawling nearby.

Algae

Fungus

The fungus part of a lichen can't make sugar for food. Only the alga part can. But as the fungus absorbs sugar, it protects the alga and helps keep it from drying.

Mixed in with the lichens are strands of green algae. Can you see them under your magnifying glass?

A filelike tongue, or radula, makes it easy for periwinkles and limpet snails to scrape algae off rocks.

Radula

Life in a Shower

If you stand at the top of your square, above the black line, there's a good chance that waves smashing against rocks below will send up an ocean spray. It splashes you as it wets the plants, animals, and rocks nearby. This saltwater shower doesn't annoy the living things around you. It helps keep them moist when the sun beats down and the wind blows.

Look closely at the rocks with your magnifying glass. That slime you see is a layer of tiny blue-green bacteria. Like plants, these bacteria use energy from the sun to make sugar from water and carbon dioxide gas.

Now look at the black line. It is mostly made of lichens. A lichen is an alga and a fungus living together. Try to find plant-eating snails such as periwinkles and limpets. As they graze, they may become a meal for a hungry shorebird or meat-eating dog-winkle snail.

Where there are plant eaters there are meat eaters. A dogwinkle can drill a hole through a periwinkle's shell, then eat the animal inside. In this case, both the predator and the prey are rocky shore snails.

Zoned for Life

Rocky seashores are divided into zones where different kinds of plants and animals are adapted to live. Growing in the splash zone are blue-green bacteria. Just below them are lichens that form a black line on the rocks. Growing in the other zones are plants that form colored bands.

List in your notebook the order of the colored bands you see in your square, starting with the black line. As you explore, draw pictures of the plants and animals that live in each zone. Then use a field guide to identify them. (See page 15.)

Remember to check tide tables every day, so you explore only when the tide is going out. (See page 7.)

X Marks the Spot

If you find a limpet on a rock, draw a chalk circle around it. Check to see if and when the limpet moves from its rest spot and if it always returns home.

25

Though seaweeds make sugar like other plants, they don't have leaves, stems, roots, or flowers. You may see a crab rip off a piece of seaweed with its pincer claw.

Holdfast   *Spores*

Look for clusters of pale-yellow egg cases stuck to rocks by dogwinkle snails.

On the upper surface of a snail's muscular foot is the operculum. When a snail senses danger or needs to trap water inside, it pulls into its shell. The operculum seals the opening.

A hungry dogwinkle drills a hole into the cone of a barnacle, forces the cone open, and reaches its prey.

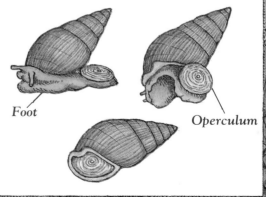

Foot   *Operculum*

You eat with your hands. But barnacles catch food with their feet. If you find a barnacle-covered shell, place it in seawater and you may see the barnacles' feet move.

Where Waves Crash

When the tide is out, cross the black line of lichens and enter the part of your square where plants and animals are able to make a living underwater, exposed to the air, and in the path of waves crashing on shore.

Feel the brown, green, and red algae drooping over the rocks. Their wetness is a clue that these plants, like the sand on a beach, offer a damp, cool, protected place for animals to hide from sun, wind, and waves until the tide returns to cover them.

The algae growing on the rocks between the tides are seaweeds—plants that are very different from those growing in a forest or in a backyard. Seaweeds don't have leaves. Instead, they have blades that don't tear easily. Instead of stems, they have flexible stalks that bend when waves break on them. And instead of roots, seaweeds grow holdfasts that anchor them to rocks so crashing waves can't pull them away from shore.

Look for seaweeds that seem to be covered with tiny bubbles. The bubbles are gas-filled air bladders that float the blades in the water. The blades absorb sunlight so the plant can make enough sugar to grow and reproduce.

Unlike other crabs' shells, the hermit crab's own shell doesn't protect its back end (left). So the hermit (below) finds an empty snail shell to do the job.

Hairy hermit crab

27

Algae Art

algae

Plants are beautiful to look at, but they can also be turned into art. Search your square for algae broken or loosened by waves. Soak each piece in water for a few minutes. Then arrange the pieces on a sheet of paper the way you want them to look when dry.

waxed paper

algae

paper

stack of books

Sandwich the sheet of paper between two pieces of waxed paper and stack some books on top. Check once a day to see if the algae has dried onto the paper. Mount the dried algae art on a board, cover with plastic wrap, and hang it up to enjoy. You can also use a field guide to identify and label your algae once you have dried them.

As the tide rises and the blades float, the animals come out of their hiding places. Crabs and starfishes inch up rocks in search of plant-eating snails and mussels that have their shells open to filter tiny plants and animals from the ocean water. Shrimps capture worms in their claws as fishes dart in and out of the shore jungle created by the floating seaweeds. Barnacles spread their cones apart, thrusting up feathery feet to wiggle back and forth and comb food from the water into their mouths. Underwater, each animal also breathes in fresh oxygen and gets rid of wastes.

The hours go by. One by one, animals of the rocky shore sense the tide turning. Soon they will be out in the air. Every wave that crashes on shore will pound them too. Barnacles pull in their feet, trap water in

The pullback of the tide signals it is time for barnacles, mussels, and other attached animals of the rocky shore to seal themselves in with enough water to breathe and stay moist.

If you find an animal with a shell made of overlapping plates, it is a chiton.

A sea slug stranded on shore searches for seaweeds to hide under until the tide returns.

In the circle are tiny water plants that shore animals eat.

28

their cones to breathe and keep moist, and close up shop. They are cemented to rocks. You can't move them, and neither can the waves.

Limpets return to their rest spots, fill up on water, and clamp down on the rocks. Just try budging one of them!

Mussels take in their last drops of water until the next high tide and shut themselves tightly into their shells. Look for the tough byssus threads that fasten mussels securely to the rocks.

As the water goes out, crabs hide under loose rocks or in cracks between rocks. Many snails and worms wait for floating seaweeds to fall gently on top of them when the water is gone.

Day after day, the in-again, out-again cycle of the tides continues. Predators may eat some of the animals in your square, but only the fury of storm waves pounding the rocks have enough power to pull attached creatures into the sea.

Pick up a starfish, turn it over, and you will see rows of tiny tubes that have suction cups for gripping rocks. These cups can pull open a clam or mussel shell, though it may take hours to do it. The cups are called tube feet.

Suction cups

Those brownish and greenish blobs on the rocks are most likely sea anemones that drew in their tentacles and trapped water inside when the tide went out. Leave them alone. Their tentacles sting!

29

Tide pools are alive with hunters and hiders seeking food and safety.

Keep a list of the birds that feed at the tide pool.

If you pull animals out of a tide pool with your dipping net, return them to the water after a few minutes. If you poke around very gently with your stick, you should do no harm.

The dwarf brittle star moves with its arms to catch worms or small shrimps.

Dwarf brittle star

The Small Square Aquarium

There's more than meets the eye in those "puddles" the tide leaves behind on a rocky shore. They are tide pools—shallow hollows in the rocks that are filled with seawater. And beneath the calm surface of nearly all tide pools, there is a world of ocean life you could not view unless someone took you scuba diving or for a ride in a glass-bottomed boat.

Approach a tide pool slowly. Sit down beside it so your shadow doesn't fall on the water. Be very still.

Scare an octopus, and it may squirt out a cloud of dark ink that will confuse you—long enough for the octopus to speed through the water to safety.

An oystercatcher can slip its beak into an open oyster shell and pull the oyster out to eat.

Most likely the animals in the pool have sensed your motion and darted for safety rather than chance being eaten by a predator. But be patient, and keep looking into the water. Those brightly colored "plants" are animals—sea anemones or peacock worms waving their tentacles to capture food. Watch for a mussel or a barnacle to open. Soon big and small fishes appear, along with other full-time tide pool residents, including crabs, snails, sea urchins, shrimps, sea slugs, and starfishes. You may even be lucky enough to catch a glimpse of a young octopus before it swims away when the tide returns.

Colorful sea slugs must not be tasty—few if any tide pool animals eat them.

Many tide pool plants look like animals, while many animals look like plants or rocks.

31

Shore in a Box

Take a shoebox and measure how long and high it is. Cut paper a little less than the height but about four inches longer. Draw and color rocks, plants, and shore animals. Place the picture in the shoebox, and tape each side to the front. The picture will curve.

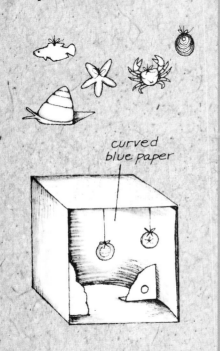

curved blue paper

Now draw more rocks, each with a flap at the bottom. Cut out each one, bend its flap, and glue it to the bottom of the box. To construct a tide pool, use blue paper for the water background. Draw and cut out fishes and hang them from the top of the box with taped string.

Little black sea cucumber

Orange lichen

Thrift

Acorn barnacle

Red lineus worm

Purple shore crab

Sea lettuce (green algae)

Black Katy chiton

Sometimes the sun evaporates too much water from a tide pool. Then the water that is left is very high in salt and very low in oxygen. Look into such a pool and you will see animals that move slowly and look tired. Some animals may die before the next high tide comes to the rescue with oxygen-rich seawater.

It's easy to spend hours gazing into the sea-life aquariums that nature provides in the tide pool on a rocky shore, especially if you are taking notes and drawing pictures in your notebook. Just remember to **WATCH THE TIDE.**

Life at the Edge

Compared to forests, deserts, and the ocean itself, the seashore may seem a small part of nature. But it is the place where you meet thousands of wild animals that can do what whales, frogs, camels, and trees can't— live both in and out of the sea every day.

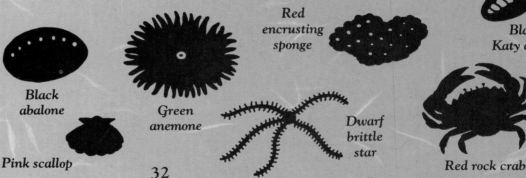

Black abalone

Pink scallop

Green anemone

Red encrusting sponge

Dwarf brittle star

Red rock crab

32

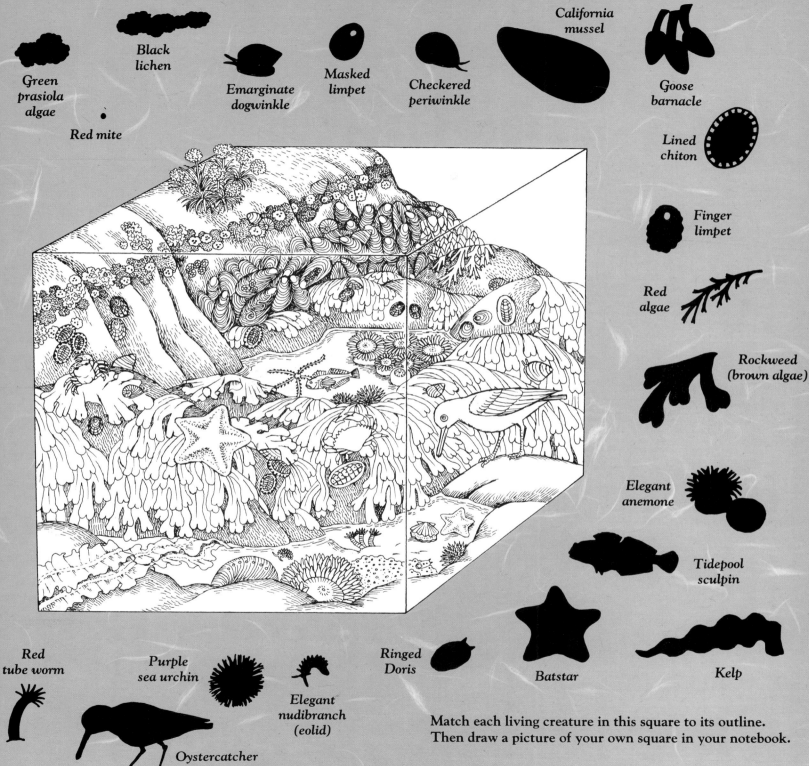

Green prasiola algae

Red mite

Black lichen

Emarginate dogwinkle

Masked limpet

Checkered periwinkle

California mussel

Goose barnacle

Lined chiton

Finger limpet

Red algae

Rockweed (brown algae)

Elegant anemone

Tidepool sculpin

Red tube worm

Purple sea urchin

Elegant nudibranch (eolid)

Oystercatcher

Ringed Doris

Batstar

Kelp

Match each living creature in this square to its outline.
Then draw a picture of your own square in your notebook.

For Keeps

Soak overnight in water the shells you want to keep. If an animal comes out, put the shell back in the ocean. When you go home, take the empty shells, soak them in warm water and dishwashing liquid, and rinse and dry them. Examine each shell with your magnifying glass, and try to identify it using a field guide. Write the name of each shell, and when and where you found it, on a card. Place that shell on top of its card in a box.

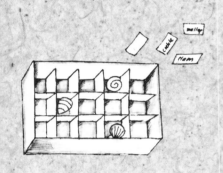

As your collection grows, group similar shells together. Start your own shell book by drawing every shell in your collection, noting its name and where you found it. Add to your book every time you visit the seashore.

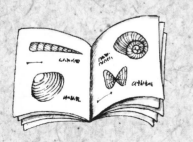

Match each living creature in this square to its outline. Then draw a picture of your own square in your notebook.

The only creatures that survive on a seashore are the ones adapted to making a living there. And even they can't exist everywhere on shore. The ones that survive out of water the longest live close to the high tide line. Those that can be out of water only a short time live close to the low tide line. In between, each creature lives where it is best fitted to find food and stand the wind, sun, waves, and predators. That's why it is so important that you never take live animals away from the seashore and that you put back any animal you pick up to examine in the place you found it.

Hundreds of millions of people around the world visit the seashore each year. With so many people, it's easy to understand why shore life is in danger. Every time oil pours out of tankers at sea and washes up on shore, animals and plants are killed, and sand and rocks are made unfit for living. Paper, cans, plastic, and other trash also harm the shore, as do chemicals and detergents that come from factories and homes. So do your best to keep the seashore pollution-free.

The more time you spend in your small square, the more you will appreciate how life at the edge of the sea works. It is, indeed, one of nature's great wonders.

Sea oats

Moon snail

Strawberry cockle

Sand shrimp

Angel wing

Surf clams

Jackknife clam

Parchment worm

Clam worm

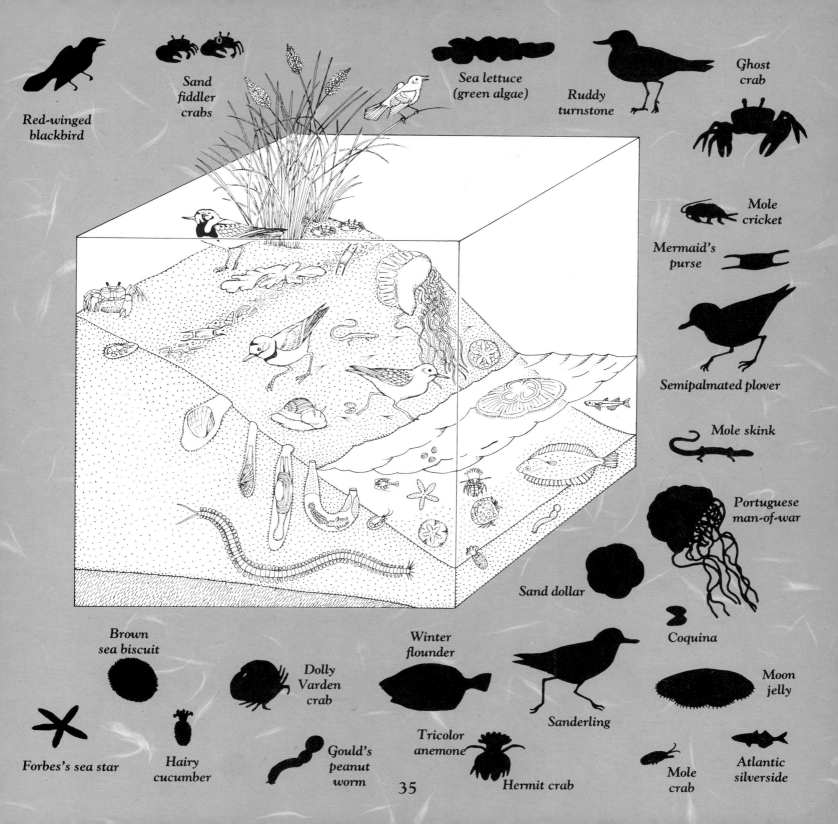

Red-winged blackbird

Sand fiddler crabs

Sea lettuce (green algae)

Ruddy turnstone

Ghost crab

Mole cricket

Mermaid's purse

Semipalmated plover

Mole skink

Portuguese man-of-war

Sand dollar

Coquina

Brown sea biscuit

Dolly Varden crab

Winter flounder

Sanderling

Moon jelly

Forbes's sea star

Hairy cucumber

Gould's peanut worm

Tricolor anemone

Hermit crab

Mole crab

Atlantic silverside

Just about all of the shells you find in your square were made by soft-bodied animals called mollusks. Snails are mollusks that live inside a coiled shell. Bivalves live inside a two-piece shell. Slugs are snails with little or no shell, and octopuses lack shells, too.

Snails

Black abalone

Black turban

Checkered periwinkle

Purple snail

Sitka periwinkle

Dove snail

Florida fighting conch

Emarginate dogwinkle

Channeled dogwinkle

Lettered olive snail

Laminated dogwinkle

Common nutmeg

Spindle shell

Lightning whelk

Boring turret snail

Eastern dog whelk

Alphabet cone

Bivalves

California mussel

Blood ark

Common mussel

Rigid penshell

Pacific pink scallop

Angel wing

Razor clam

Jackknife clam

Turkey wing

False Pacific jingle shell

Astarte clam

Pennsylvania lucine

Strawberry cockle

Dwarf surf clam

Egg cockle

Coquina

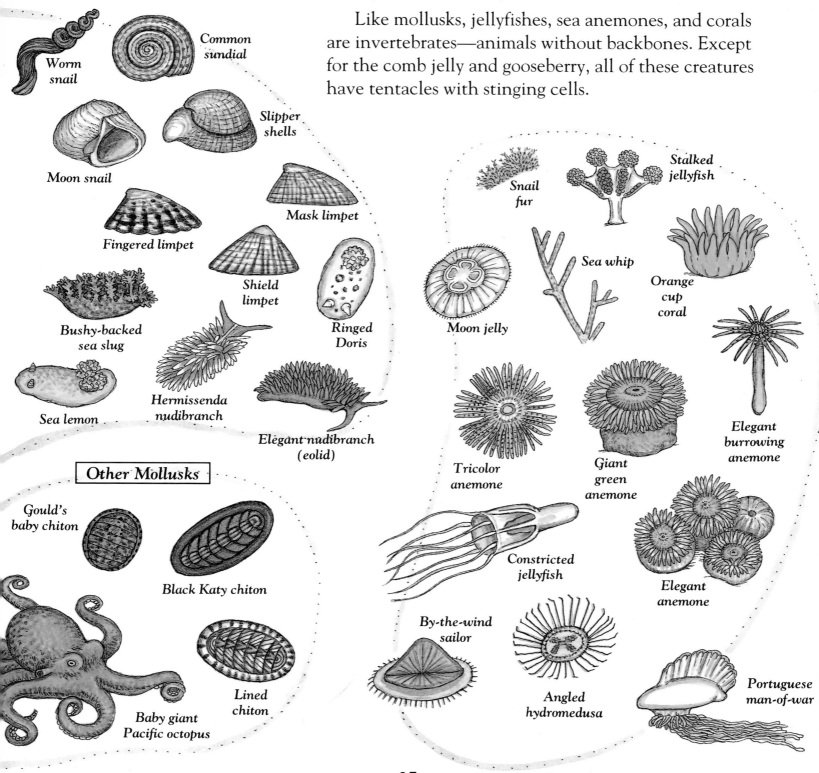

Worm snail

Common sundial

Moon snail

Slipper shells

Fingered limpet

Mask limpet

Shield limpet

Bushy-backed sea slug

Ringed Doris

Sea lemon

Hermissenda nudibranch

Elegant nudibranch (eolid)

Like mollusks, jellyfishes, sea anemones, and corals are invertebrates—animals without backbones. Except for the comb jelly and gooseberry, all of these creatures have tentacles with stinging cells.

Snail fur

Stalked jellyfish

Sea whip

Moon jelly

Orange cup coral

Tricolor anemone

Giant green anemone

Elegant burrowing anemone

Other Mollusks

Gould's baby chiton

Black Katy chiton

Baby giant Pacific octopus

Lined chiton

Constricted jellyfish

Elegant anemone

By-the-wind sailor

Angled hydromedusa

Portuguese man-of-war

37

All these shore creatures are boneless echino-
derms—spiny-skinned animals. They depend on their
tough skin, plates, and spines or bumps for protection
from hungry predators.

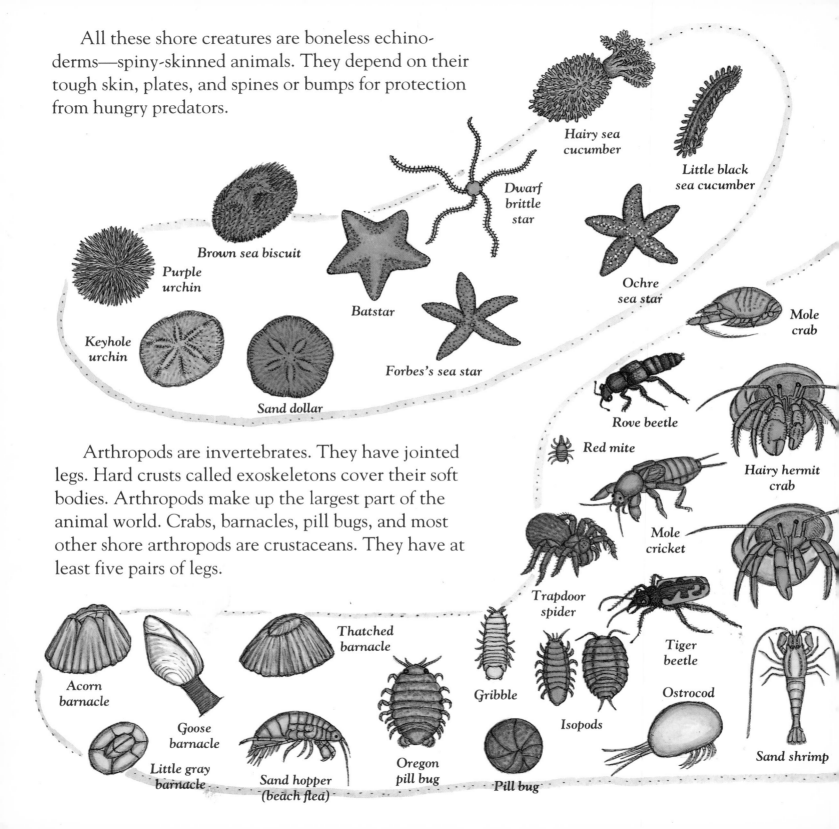

Hairy sea
cucumber

Little black
sea cucumber

Dwarf
brittle
star

Brown sea biscuit

Purple
urchin

Ochre
sea star

Keyhole
urchin

Batstar

Mole
crab

Sand dollar

Forbes's sea star

Arthropods are invertebrates. They have jointed
legs. Hard crusts called exoskeletons cover their soft
bodies. Arthropods make up the largest part of the
animal world. Crabs, barnacles, pill bugs, and most
other shore arthropods are crustaceans. They have at
least five pairs of legs.

Rove beetle

Red mite

Hairy hermit
crab

Mole
cricket

Trapdoor
spider

Tiger
beetle

Thatched
barnacle

Acorn
barnacle

Gribble

Ostrocod

Goose
barnacle

Isopods

Little gray
barnacle

Sand hopper
(beach flea)

Oregon
pill bug

Pill bug

Sand shrimp

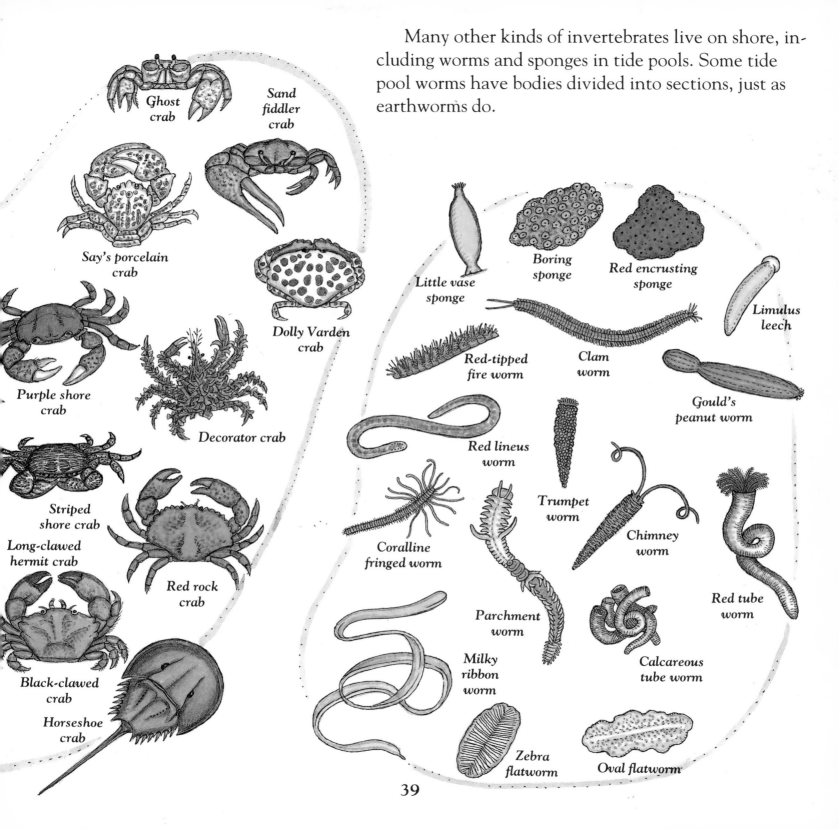

Many other kinds of invertebrates live on shore, including worms and sponges in tide pools. Some tide pool worms have bodies divided into sections, just as earthworms do.

Ghost crab

Sand fiddler crab

Say's porcelain crab

Dolly Varden crab

Purple shore crab

Decorator crab

Striped shore crab

Long-clawed hermit crab

Red rock crab

Black-clawed crab

Horseshoe crab

Little vase sponge

Boring sponge

Red encrusting sponge

Limulus leech

Red-tipped fire worm

Clam worm

Gould's peanut worm

Red lineus worm

Trumpet worm

Coralline fringed worm

Chimney worm

Parchment worm

Red tube worm

Milky ribbon worm

Calcareous tube worm

Zebra flatworm

Oval flatworm

39

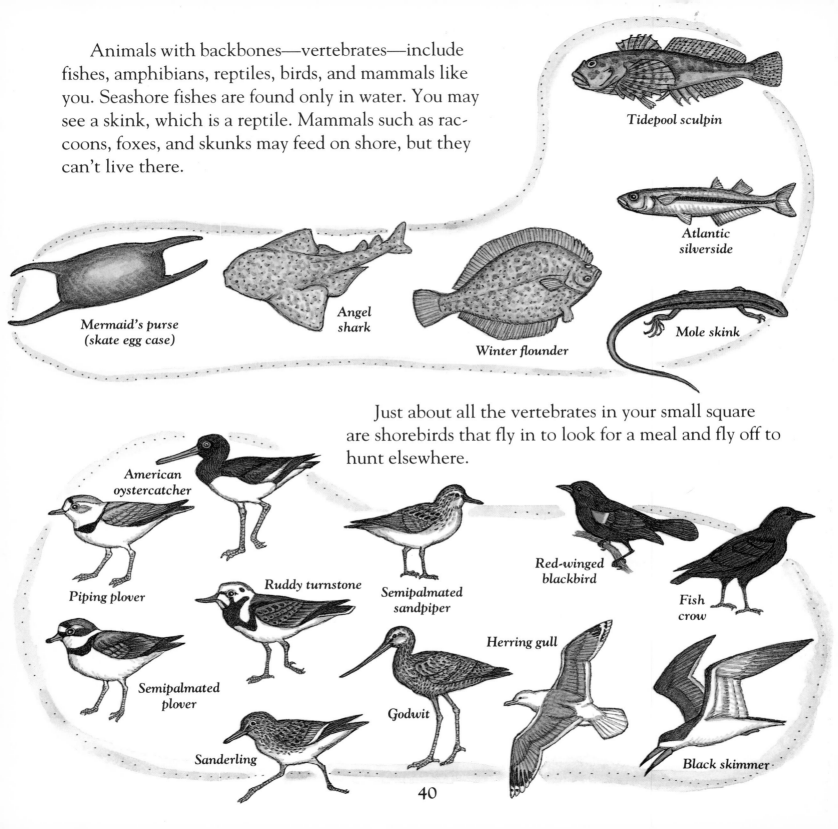

Animals with backbones—vertebrates—include fishes, amphibians, reptiles, birds, and mammals like you. Seashore fishes are found only in water. You may see a skink, which is a reptile. Mammals such as raccoons, foxes, and skunks may feed on shore, but they can't live there.

Tidepool sculpin

Atlantic silverside

Mermaid's purse (skate egg case)

Angel shark

Winter flounder

Mole skink

Just about all the vertebrates in your small square are shorebirds that fly in to look for a meal and fly off to hunt elsewhere.

American oystercatcher

Piping plover

Semipalmated plover

Ruddy turnstone

Semipalmated sandpiper

Red-winged blackbird

Fish crow

Sanderling

Godwit

Herring gull

Black skimmer

40

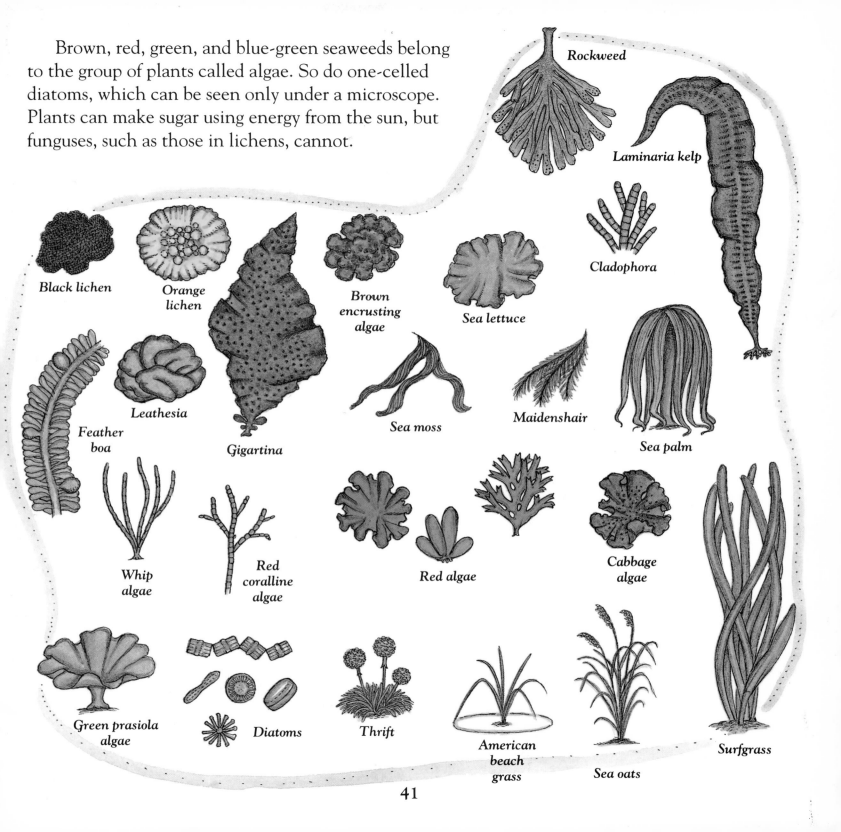

Brown, red, green, and blue-green seaweeds belong to the group of plants called algae. So do one-celled diatoms, which can be seen only under a microscope. Plants can make sugar using energy from the sun, but funguses, such as those in lichens, cannot.

Rockweed

Laminaria kelp

Cladophora

Black lichen

Orange lichen

Brown encrusting algae

Sea lettuce

Feather boa

Leathesia

Gigartina

Sea moss

Maidenshair

Sea palm

Whip algae

Red coralline algae

Red algae

Cabbage algae

Green prasiola algae

Diatoms

Thrift

American beach grass

Sea oats

Surfgrass

41

If you are lucky, you may spot a big animal, a really big animal, or a rare animal on the shore or in the water.

American manatee

Sea otter

Off the Florida coast you may see a manatee pop its head out of the water to breathe. Off the California coast look for a sea otter floating on its back. It may be eating a clam or other shelled animal it caught. Both manatees and sea otters are mammals and both are endangered.

Shark　　　*Dolphin*

A fin sticking out of the water may belong to a shark or a dolphin. Dolphins are fast-swimming whales. You may spot one leaping out of the water.

Sea lion

Seal

Seal　　　*Sea lion*

Seals and sea lions are mammals. They spend part of their lives in water and part on land. Sea lions have outer ears but seals do not. A sea lion can turn its back flippers forward. A seal cannot.

Almost all anyone ever sees of a whale is its tail or spout. The spout is water gas in the whale's breath that turns to water drops in the cooler air. A whale's nose is on top of its head. Because a whale is a mammal, it must surface to breathe.

Humpback whale spout

Sperm whale spout

Humpback whale

Gray whale

If you find sea turtle eggs, do not disturb them. Sea turtles are endangered animals.

Eggs

Sperm whale

Every year this sea turtle visits the same beach. She comes ashore and digs a pit in the sand. Then she lays eggs, covers them, and returns to the sea.

Kemp's Ridley sea turtle

43

Index

A

adaptation 11, 25, 34. *Any part of a living thing that makes it fitted to survive where it lives.*

air bladder 27

alga (AL-guh) *plural* **algae** (AL-jee) 24, 25, 27, 41

amphibian (am-FIB-ee-in) 40. *Bony animal that lives the first part of its life in water and the second part on land.*

arthropod (AHR-thruh-pahd) 38

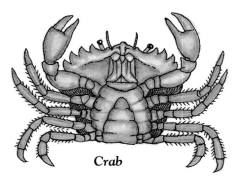

Crab

Spider

Barnacle

B

bacteria 25. *Kinds of monera— one-celled creatures that don't have a nucleus.*

barnacle (BAHR-nuh-kill) 12, 21, 23, 26, 28, 31, 38

beach 3-19, 21, 34-35

beach grass 10

beak 11, 16, 19, 31

beetle 8, 12, 13, 15

bill (*see* **beak**)

bird 3, 5, 10, 12, 13, 15, 16, 17, 19, 30, 40

bivalve 36

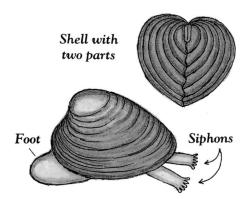

Shell with two parts

Foot   *Siphons*

blackbird 10

blade 27, 28

boulder 21

brittle star 30

burrow 11

byssus (BISS-iss) 17, 29

by-the-wind sailor 19

C

camouflage (KAM-uh-flahj) 18. *An animal's color, pattern, or shape that helps it hide in its surroundings from other animals that want to eat it.*

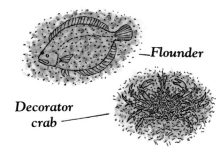

Flounder

Decorator crab

carbon dioxide 25. *Gas found in the air that plants use to make food.*

cell 16. *Smallest living part of all plants, animals, and funguses. Some living things are made up of just one cell.*

chiton (KYT-in) 28

clam 3, 8, 12, 13, 16, 17, 42

claw 26, 28

cliff 3, 7, 21

comb jelly 37

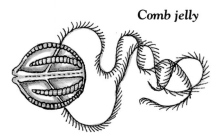

Comb jelly

copepod (KOH-puh-pahd) 16

coral 13, 37

Single polyp

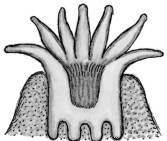

Colony

Index

crab 3, 9, 10, 11, 12, 18, 19, 26,
    27, 28, 29, 31, 38
cricket 13
crustacean (kris-TAY-shin) 38

D
diatom (DY-uh-tahm) 41
dogwinkle 25, 26
dolphin 42
driftwood 8, 13

E
earwig 12
echinoderm (uh-KY-nuh-durm)
    38
egg 14, 15, 43. *To scientists, an
egg is a female reproductive
cell. The eggs you find on the
seashore may have young
growing inside.*
endangered 42, 43
energy 5, 6, 25, 41. *Ability to do
work or to cause changes in
matter.*
erosion 6, 10
evaporation 32
exoskeleton 38

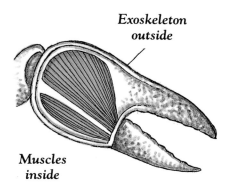

*Exoskeleton
outside*

*Muscles
inside*

F
field guide 15
fish 13, 18, 19, 23, 28, 31, 40
flatworm 12, 14
flounder 18
foot 8, 23, 26, 28, 29
footprint 10
fox 10, 40
fungus 11, 24, 25, 41

G
ghost crab 10, 11
gills 14. *Breathing parts of fish and
many other water animals.*

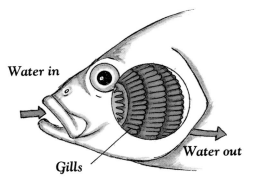

Water in

Water out

Gills

gooseberry 37
gravity 6, 7. *Force that pulls one
object toward another.*
gribble 13
gull 6, 13, 15

H
hermit crab 12, 27
holdfast 27
horseshoe crab 14

I
igneous rock 7. *Rock formed
when lava cools and hardens.*
ink 31

intertidal zone 6. *The part of the
seashore covered by water at
high tide but exposed to air at
low tide.*

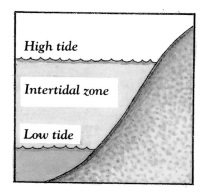

High tide

Intertidal zone

Low tide

invertebrate 37, 39. *Animal
without a backbone.*

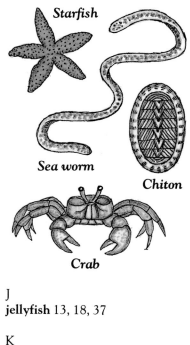

Starfish

Sea worm

Chiton

Crab

J
jellyfish 13, 18, 37

K
krill 16

45

Index

L
leech 14
lichen (LY-kin) 24, 25, 27, 41
limpet 22, 24, 25, 29

M
magnifying glass 5, 7, 11, 12, 15, 21, 24, 25
mammal 41, 42
manatee 42
mermaid's purse 13

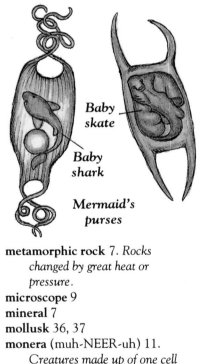

Baby skate

Baby shark

Mermaid's purses

metamorphic rock 7. *Rocks changed by great heat or pressure.*
microscope 9
mineral 7
mollusk 36, 37
monera (muh-NEER-uh) 11. *Creatures made up of one cell that doesn't have a nucleus.*
moon 6, 7
moon snail 15, 17
mucus (MYOO-kiss) 15
mussel 22, 28, 29, 31

N
nest 10

O
ocean 6
octopus 31, 36

operculum (oh-PUR-kyuh-lim) 26
ostracod 8
oxygen (AHK-suh-jin) 14, 17, 28, 32. *A gas that living things breathe. Oxygen is in the air and dissolved in seawater.*
oyster 31
oystercatcher 31

P
pebble 21
pelican 19
periwinkle 24, 25
pill bug 10, 11, 15, 38
plankton 9
plover 10, 12, 13
pollution 13, 34
Portuguese man-of-war 13
predator (PRED-uh-tur) 11, 12, 17, 18, 22, 29, 31, 38. *Animal that kills other animals for food.*
prey 18
protist 11. *Creature usually of one cell that has a nucleus.*

Q
quartz 7

R
raccoon 10, 13, 40
radula (RAJ-uh-luh) 24
reptile 40. *Bony animal with dry skin covered by scales or plates.*
rock 7, 14, 21, 23, 24, 27, 29, 30

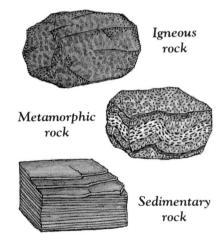

Igneous rock

Metamorphic rock

Sedimentary rock

rocky shore 3, 8, 19, 20-33
root 10

S
salt 32
sand 3, 7, 8, 10, 11, 14, 15, 16, 17, 18, 21, 27
sand dollar 14, 15, 18
sanderling 14
sand hopper 8, 10, 11
sandy shore (see beach)
scallop 19
sculpin 23
sea 3, 8, 18, 19
sea anemone (uh-NEM-uh-nee) 12, 22, 29, 31, 37

Index

sea biscuit 18
seal 42
sea lion 42
sea otter 42
sea slug 22, 28, 31, 36
sea turtle 43
sea urchin 13, 31
seaweed 3, 5, 8, 11, 13, 26, 27, 28, 29, 41
sedimentary rock 7. *Rock that forms from pieces of other rocks or shells.*
shark 13, 18, 19, 42
shell 3, 5, 8, 11, 12, 13, 14, 15, 16, 19, 22, 25, 26, 27, 28, 29, 31, 34, 36
shipworm 13
shorebird 12, 25, 40
shoreline 7
shrimp 18, 28, 30, 31
skate 13

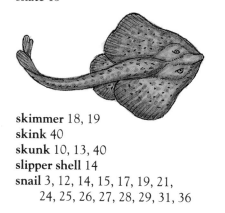

skimmer 18, 19
skink 40
skunk 10, 13, 40
slipper shell 14
snail 3, 12, 14, 15, 17, 19, 21, 24, 25, 26, 27, 28, 29, 31, 36

soil 24
sponge 22, 39

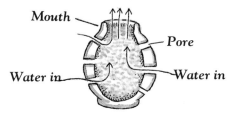

Mouth — Pore — Water in — Water in

spout 43
starfish 14, 15, 18, 19, 28, 29, 31
stinging cell 18, 37
strandline 8, 9, 10, 11, 12, 13, 14, 15, 22
sugar 24, 25, 26, 27, 41
sun 16, 21, 22, 25, 27, 32, 34, 41
surf 6

T
tentacle 18, 22, 29, 31, 37
thrift 24
tide 3, 6, 7, 8, 12, 13, 14, 15, 16, 17, 19, 21, 22, 23, 24, 27, 28, 34
tide pool 22, 23, 30-32, 39
tide table 7, 25
tool 5
track 10
tube 15, 16, 17
tube foot 29
tunnel 16
turnstone 11

V
vertebrate 40. *Animal that has a backbone.*

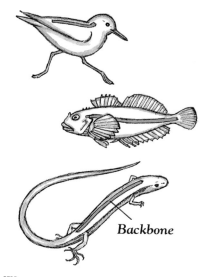

Backbone

W
water 3, 5, 6, 7, 12, 13, 14, 16, 17, 18, 19, 21, 22, 25, 27, 28, 29, 30, 32, 34
wave 3, 5, 6, 7, 19, 21, 22, 25, 27, 29, 34
whale 32, 43
wind 5, 11, 21, 22, 34
wood 13, 19
woodlice 13
worm 3, 12, 13, 16, 17, 19, 28, 29, 30, 31, 39

Everywhere the sea meets the land there is a sandy or a rocky seashore. This map shows where the animals and plants in this book can be found.

Further Reading

To find out more, look for the following in a library or bookstore:

Golden Guides, Golden Press, New York, NY

Golden Field Guides, Golden Press, New York, NY

The Audubon Society Beginner Guides, Random House, New York, NY

The Audubon Society Field Guides, Alfred A. Knopf, New York, NY

The Peterson Field Guides, Houghton Mifflin Co., Boston, MA

Reader's Digest North American Wildlife, Reader's Digest, Pleasantville, NY

Look in an art supply store or in the library for books on how to draw plants and animals. If you like to sketch and paint outdoors, here are some things you'll find handy:

- paper
- number 2 pencil
- paintbrush
- bottle of black ink
- tray of watercolors
- eraser
- plastic bottle for water
- stiff cardboard or clipboard to draw on

DATE			